Clemens Herschel, American Society of Civil Engineers

The Venturi Meter

An Instrument Making Use of a New Method of Gauging Water

Clemens Herschel, American Society of Civil Engineers

The Venturi Meter
An Instrument Making Use of a New Method of Gauging Water

ISBN/EAN: 9783337140199

Printed in Europe, USA, Canada, Australia, Japan

Cover: Foto ©berggeist007 / pixelio.de

More available books at **www.hansebooks.com**

The Venturi Meter

AN INSTRUMENT MAKING USE OF A NEW METHOD OF
GAUGING WATER; APPLICABLE TO THE CASES OF
VERY LARGE TUBES, AND OF A SMALL VALUE
ONLY, OF THE LIQUID TO BE GAUGED.

BY

CLEMENS HERSCHEL, M. Am. Soc. C. E.

Read before the American Society of Civil Engineers,
December 21, 1887.

———————

For this paper the American Society of Civil Engineers awarded
Mr. Herschel the Rowland Prize.

———————

REPRINTED BY

BUILDERS IRON FOUNDRY,

PROVIDENCE, R. I.

1898

The Venturi Meter

AN INSTRUMENT MAKING USE OF A NEW METHOD OF
GAUGING WATER ; APPLICABLE TO THE CASES OF
VERY LARGE TUBES, AND OF A SMALL VALUE
ONLY, OF THE LIQUID TO BE GAUGED.

By Clemens Herschel, M. Am. Soc. C. E.

Read Before the American Society of Civil Engineers
December 21st, 1887.

"Introduction; analogy; assumptions founded on facts and unceasingly
rectified by additional observations: a genial form of tact, inborn, but
strengthening itself by making numerous comparisons between its
indications and the results of experiment; such are the principal
means of arriving at the truth."—Laplace, *des divers moyens d'approcher
de la certitude.*

New Meter
needed.

Some additional instrument or method for gauging
water has long been desired by hydraulic engineers. In
the case of water flowing through pipes, as in city water-
works, it is extremely difficult or impracticable to meter
the water, as soon as diameters approaching one foot, or
quantities approaching one million gallons daily, are
reached. In some such cases the stream of water has
been split up into many smaller ones, each of which was
then furnished with a meter, and the tail water of these
meters reunited — a method and apparatus so cumber-

some and costly as to be rarely applicable. Taking the case, on the other hand, of a far less valuable commodity, viz., of water under little or no pressure, about to be or after it has been used for power, the practical difficulties of gauging again become very great. Ordinary meters are out of the question, owing to the small value of the article per cubic foot, and to the proportionately great cost, per cubic foot of water metered, of applying a mechanical meter.   *   *   *

It has long seemed to the writer that an application to metering water of the principle involved in the Bourdon anemometer, an instrument which has been used to measure the velocity of currents of air in mines, in France, would yield valuable results, and the present paper is intended to record the experiments made and the results found with two sizes of water meter of that description. Bourdon's anemometer is founded upon the property of a Venturi tube to exercise a sucking action through holes bored into its narrowest section. Then by measuring the intensity of this aspiration by means of any form of vacuum gauge, and establishing the relation between such "vacuum pressure" and the velocity of the air through the tube, the instrument becomes an anemometer.

Application of the principle in Bourdon Anemometer to measuring water.

This described property of the Venturi tube was known to Venturi, and may be found detailed at length in the account of his experiments made in Modena about 1791. His own account of these experiments was published in Paris in 1797, under the title "Récherches Expérimentales sur le Principe de la Communication Latérale du Mouvement dans les Fluids."* But Venturi made or suggested no use of this property, and with him it was merely a curious feature in the working of his apparatus.   *   *   *

---

* See Tracts on Hydraulics, by Thomas Tredgold, second edition, London, 1836; or Nicholson's "Journal of Natural Philosophy," Vol. III, London, 1802, for English translations; Gilbert's "Annalen," Vol. II and Vol. III, contains a German translation.

Sizes and character of Meters used for Tests.

The experiments about to be detailed were made with two sizes of Venturi Water Meters, of precisely similar interior geometric dimensions; one inserted into a tube of about nine feet, the other into a tube of about one foot in diameter. In each case the other intended dimensions may be found from an examination of the proportional dimensions given in Plate XXXIII.

The throat, or the narrowest section of the whole apparatus, is a cylinder 1 high or long, and 3 in diameter. At the distance of 1 either way from the throat, are attached the frustrums of two cones, but the angles at which the cones would meet the cylinder are rounded off; in case of the up-stream cone, on a radius of 10.38; in case of the down-stream and longer cone (the Venturi mouth-piece), on a radius of 45.83. These figures are got from making the tangents of the rounded off portions in each case=1.; the angles at the bases of the short and long cones being $79\frac{1}{2}$ and $87\frac{1}{2}$ degrees respectively. The cones are produced in each case until their diameter =9.; making the lengths from throat to end of cone 17.09 and 69.80, respectively, and the length of the whole apparatus 87.89. These are the intended dimensions, in feet, of the larger apparatus experimented with October (5–8), and by dividing all these figures by 9., we get the intended dimensions in feet, of the smaller Venturi Water Meter, which was tested June (9–15), 1887.

Description of the 12-inch Meter.

Confining ourselves now to a consideration of this last named smaller apparatus, the meter itself is shown in Plate XXXIV. The throat of the venturi, as it was named, is made of cast-iron, lined with brass, and comprises the central cylindrical and adjacent two curved portions of Plate XXXIII. Its total length was 0.563 feet. The brass lining was about $\frac{1}{2}$ an inch thick, firmly set in its envelope of cast-iron and the joint between the two end faces, cut out in form of a dove-tailed circular slot, which was then filled with Babbitt metal to guard against a possible leakage of air through the joint.

Encircling the interior narrowest section is the air-chamber, which is connected with the interior with 4 accurately and carefully drilled holes, at right angles to the center line of the venturi, and about ¼ inch in diameter each. The interior of the venturi was carefully polished with emery dust after the holes were drilled, making the edges of the 4 holes, as finally left and used, perfectly square and sharp.

To measure accurately the area of the venturi, I had made a brass cylinder, which exactly fitted it, when both were of the same temperature. This cylinder I then measured, on 3 diameters, with a vernier caliper made by Darling, Brown & Sharpe, of Providence, R. I.

The averages in each case of three such measurements, when the plug had been standing all day in the air at a temperature of about 68 degrees, after a ½-hour's immersion in ice-water, and immediately after taking it out of water of a temperature of 160 degrees Fahr., were as follows:

At 36 degrees Fahr. 3.978 inches ⎫ with no single measure-
    68   "   "   3.979   "   ⎬   ment positive as to the
    160   "   "   3.980   "   ⎭   final 0.001 inch.

From which I computed the area, at about 60 degrees Fahr., to be 0.08634 square feet, and took this as constant in all the experiments.

The air-chamber is bored at the top to receive the suction pipe, to which may be attached any form of vacuum gauge.   *   *   *

To either end of the central cast-iron member, the venturi, were attached two wooden cones, of the general interior dimensions already stated.

These were made of white pine staves, originally, or in the rough, about 2 inches thick, hooped with stout cast-iron hoops, carefully planed and scraped to smoothness inside, then soaked in water before using. The actual dimensions of the cones, measured after soaking in water, and again after the experiments, were as follows: giving the average of all the measurements taken; the differences between the first and second set being in no single measurement over 2 or 3 thousandths of a foot, and therefore insignificant.

> Smaller cone, length, 1.677, diameters, .372 and .991 feet.
> Larger cone, " 7.366, " .334 and .992 "

For purposes of the experiments, the Venturi Water Meter thus formed of the two cones and the venturi, was inserted in line of a wooden tube, made, and treated before using, as just described for the case of the two cones. The up-stream length of tube had the following dimensions:

> Length............5.007; diameter at up-stream end, .990,
> at down-stream end, .992.
> Down-stream tube.
> Length .... 5.996; diameter .... .996, .998

One foot from that end of these tubes which was joined to either of the two cones, each tube was bored to receive a piece of galvanized iron pipe, ¾-inch inside diameter. To bore these holes, a plug of soft wood was carefully fitted into the tube, under the place where the hole was to be bored. Then by using a center-bit auger, the hole could be cut through without roughing up the inside edges of the hole, and this hole left in proper shape for serving as the inside orifice of a piezometer.

The iron pipe spoken of was screwed into this hole from the outside, but was not allowed to penetrate more than about half way into the thickness of the wooden stave, .162 feet thick, forming the tube at that point.

To feed the water to the up-stream tube, without loss of head at entrance, it was furnished with a cycloidal mouth-piece, likewise made of wooden staves, carefully smoothed and soaked in water as were the other members.

This mouth-piece had a diameter of 1.001 at the outlet, 2.50 feet at the inlet end, and was 1.17 feet long; its cycloidal generator would itself be generated by a point on a circle of 0.75 feet diameter.

The experiments were conducted in the wheel-pit of the testing flume of the Holyoke Water Power Company, a ground plan of which is shown in Plate XXXV. This is a building used by the company named for testing turbines, both for purposes of the water-measurements necessary in the conduct of its own affairs, and for the public. Its foundation masonry is first class and well grouted rubble, afterwards plastered with cement, and lined with brick laid in cement. The wheel-pit end-wall, built in this same manner, is absolutely water-tight under 20 feet head of water, and it is believed that all the other walls are equally firm and water-tight. The floor consists of matched 4-inch hemlock plank, spiked to timbers resting on rows of piles, and having 2.5 foot bearing, center to center, under the wheel-pit; 4-foot bearing under the tail-race. On this first flooring is spiked another, of 4-inch matched hard pine plank under the wheel-pit; of 2-inch matched white pine under the tail-race.

Place where. Tests were made.

These statements will give a fair idea of the character of the structure which was used as a measuring tank in these experiments. For this purpose the brick walls were again plastered over with cement to give a smooth surface to measure from. The walls were then marked and divided off into rectangles by horizontal lines 1 foot apart, and by vertical lines 2 feet apart, and all dimensions carefully taken; while the floor was leveled on, both when the pit and tail-race were empty, and when full of water. I will not go into further details relating to the determination of the volume contained in the masonry tank between any two water-surfaces; nor into the determination of leakages, well known to be one of the most troublesome and laborious parts of the conduct of hydraulic investigations. Suffice it to say, that every thought of refinement of measurement and of computation was applied to the determination of volumes, while the accompanying leakages were measured at the beginning and ending of every experiment by noting the rate of rise or fall of the water surface in the tank. This water surface being generally below the level of the adjacent lower level canal, the resultant leakage was sometimes in, sometimes out, and sometimes zero.

It was never over 0.11 cubic foot per second.

The heights of water in the tank were measured by two hook-gauges, having, together, a range of about 4 feet in height, and the area of the measuring tank was about 1,150 square feet.

Description of Plates of experimental apparatus and method of conducting Tests. The whole experimental apparatus is shown, in ground plan, in Plate XXXV; in elevation, in Plate XXXVI. The water entering through the gate A, passed through several temporary divisions put into the forebay B, to quiet it, then was fed to the tube and Venturi Meter by the cycloidal mouth-piece above referred to. It was discharged into the tank C, whence it flowed into either the box D, leading through the waste-pipe E, to a by-pass F, or else, on swinging back the spout G, it was discharged into the measuring tank below. P and P^1 are piezometers, the difference of their readings indicating the head acting on the whole meter; or loss of head, caused by it. S is the suction end of the vacuum tube V, in these experiments, dipped into a tub of water, and the whole tube being 34.5 feet long, was long enough to measure a perfect vacuum, had such a thing been attainable.

This last named part of the tube was of course made of glass, in 5 lengths and with rubber-tube joints. Great care was taken to make all the joints air-tight, and by means of wrapping them with telegrapher's rubber-tape, and the use of rubber cement, this was, it was believed, successfully accomplished. * * *

A drip-box *H* caught the leakage of the spout *G*, while the water was wasting, and discharged this leakage into the waste-pipe *E* by means of the pipe *I*. *J* is a waste-valve leading to the by-pass *F*, which helped to regulate the height of water in the forebay *B*.

At the moment of one assistant opening the swing-spout *G* to discharge the water into the measuring tank, another assistant shut off the pipe *I* by means of an ordinary pipe valve next to the drip-box *H*, and he opened it again, when the swing-spout was swung back into contact with the tank *C*. The contents of the drip-box *H*, which wasted into the waste-pipe *E* at the close of each experiment, when they should have gone into the measuring tank, were added to the volume found in the measuring tank.

The times when the swing-spout was opened and when shut, were taken by the writer, with a stop-watch, reading to $\frac{1}{4}$ seconds.

The practice of the first three or four experiments sufficed to get this time, as near as he could judge, exactly right. The swing-spout was handled so easily, by means of the long lever attached, that its motion was very quick and positive. Inside of the tank *C*, and in front of the tube discharging into it, was a sliding gate, by means of which the discharge of the tube, and of the Venturi Meter, could be regulated.

It will be noted that the discharge took place under water; or, as it is generally expressed, the whole apparatus was submerged. At a later stage of the experiments, when it became desirable to reduce the amount of submergence more than was permitted by the position of the swing-spout, several experiments were made with the discharge escaping from tank *C*, through a series of holes bored into it; this discharge being measured, by comparison with other discharges of the whole system of tube, Venturi Meter and tube, when acting under the same total heads.

This series consists of experiments (70-76).   * *

Another odd set of experiments is the series (56-60). To thoroughly explain this set it may be best to speak now of the actual operations of the Venturi Meter as applied to a water-pipe in ordinary service.

If we suppose the water in the pipe to be still, the height of water in a piezometer placed just up-stream from the meter, and in the one formed by the suction pipe which leads out of the Venturi air-chamber, will be on the same level. When the water begins to flow through the Venturi it will cause the piezometric column leading out of the Venturi to fall below the straight line, joining the surfaces of the water in the two piezometric tubes placed, the one just up-stream, and the other just down-stream from the Venturi Meter; this straight line being the best obtainable reference line at the time the experiments were being conducted.

This increment of fall I have called the " depression " at the Venturi. And in the series (56-60) this depression was seen through a galvanized iron pipe by "the eye of faith " alone. Its value was taken from the values found for such depression in other experiments, when the water was passing through the Venturi with similar velocities, and the degree of submergence was such that the depression could be measured in form of a vacuum and by the vacuum gauge. In the subsequent set, the October experiments, the suction tube referred to was of glass, and the depression could be directly measured.

To resume a description of the action of the whole apparatus: As the depression increases, there comes a time when the water level in the suction tube will fall to the level of the top of the air-chamber, then fall still further, until finally it touches or blends itself with the

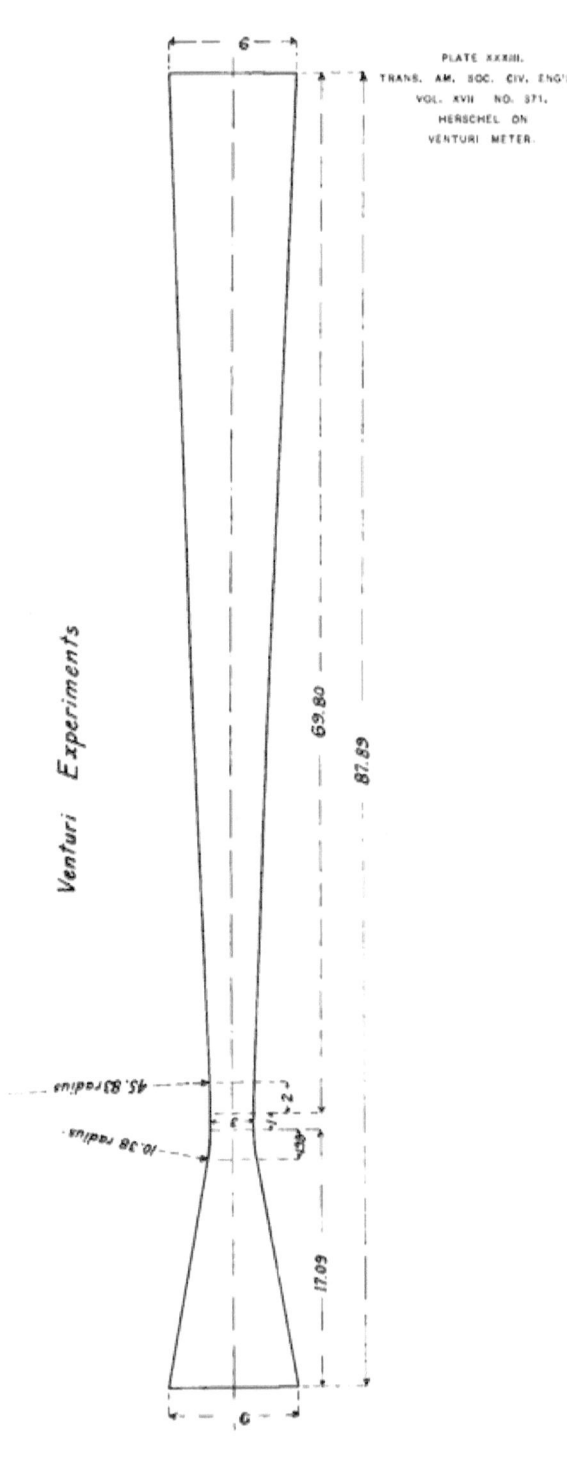

PLATE XXXIII.
TRANS. AM. SOC. CIV. ENG'RS.
VOL. XVII. NO. 371.
HERSCHEL ON
VENTURI METER.

Venturi Experiments

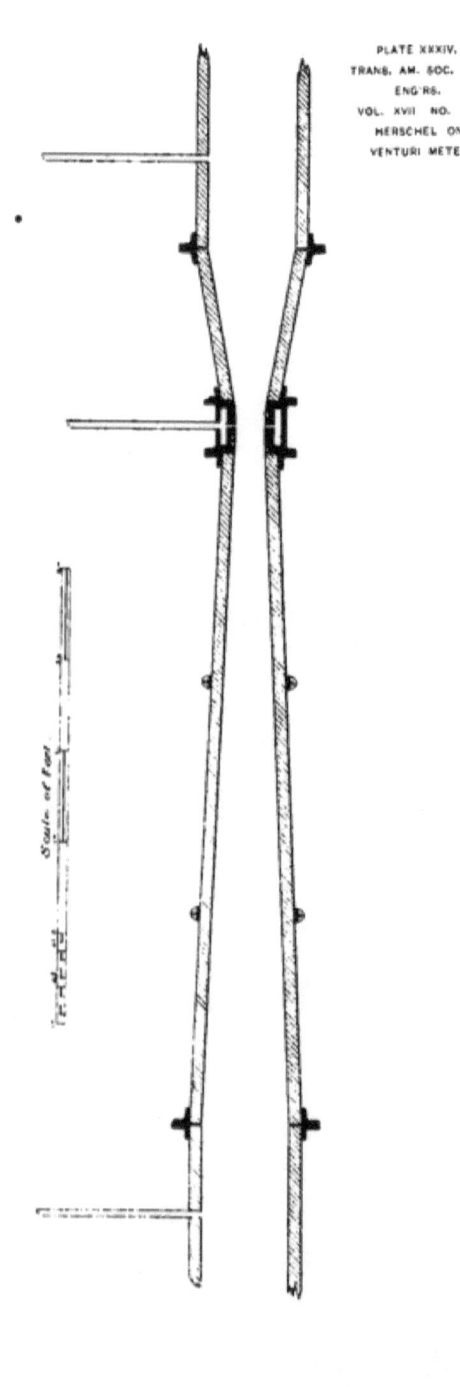

PLATE XXXIV.

TRANS. AM. SOC. CIV.
ENG'RS.

VOL. XVII   NO. 371

HERSCHEL ON
VENTURI METER

Scale of Feet

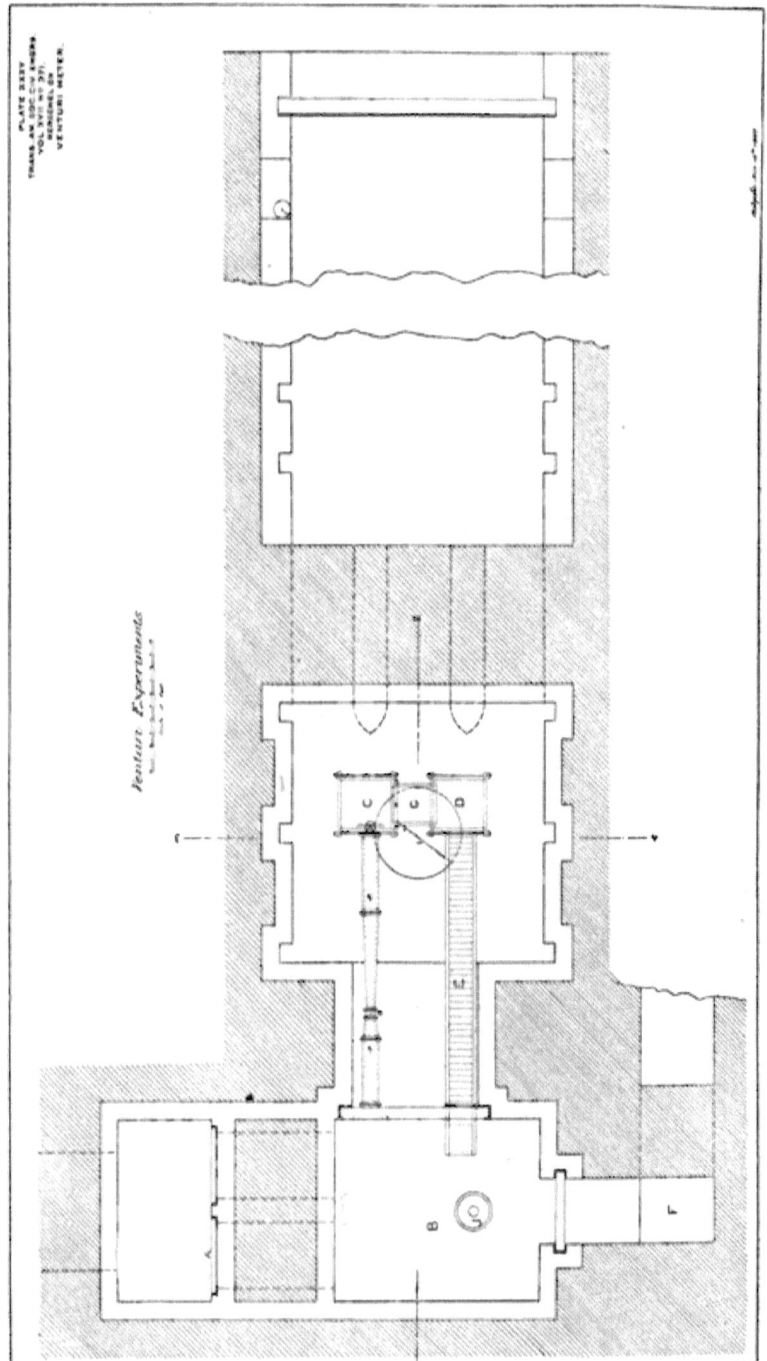

Venturi Experiments

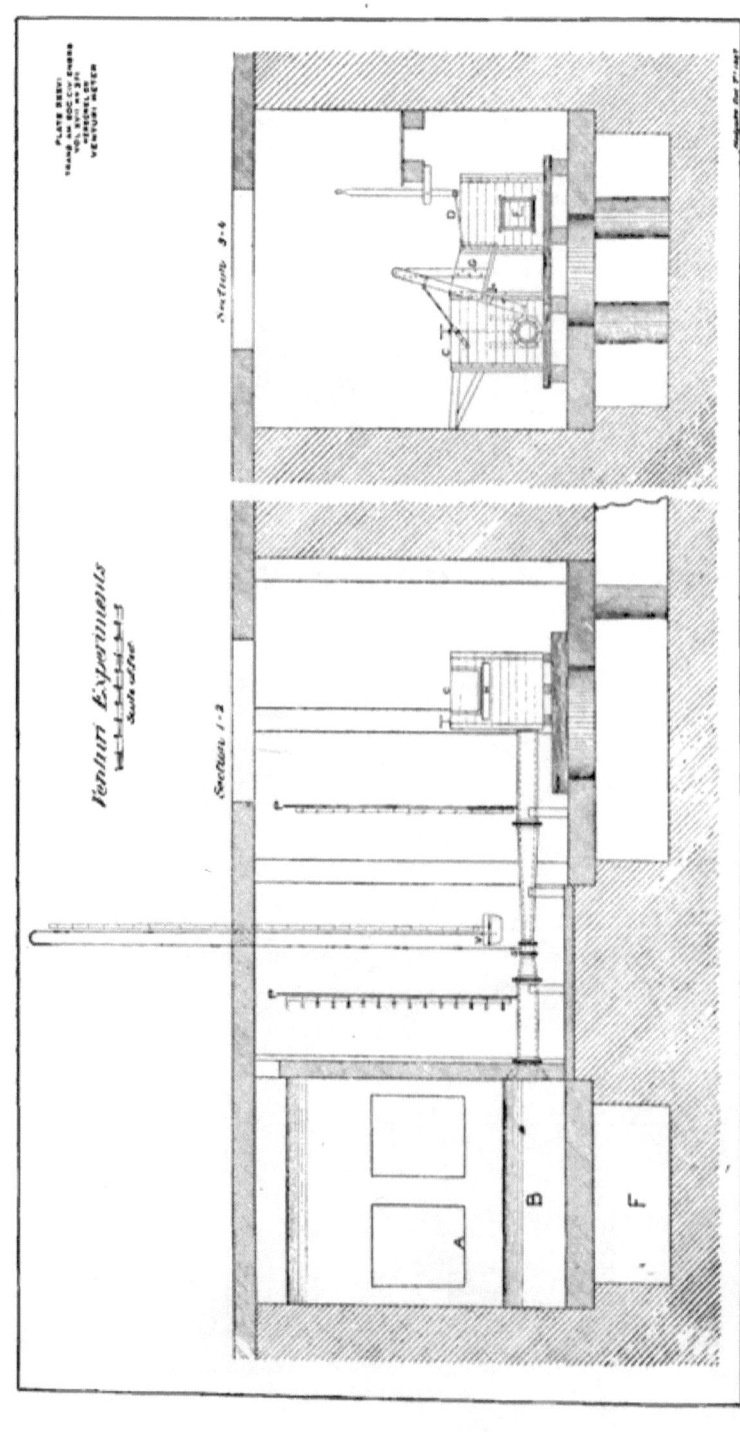

Venturi Experiments

Section 1-2

Section 3-4

PLATE XXVIII
TRANS. AM. SOC. CIV. ENGRS.
VOL. XVII, No. 327.
HERSCHEL.
VENTURI METER

surface of the stream of water spouting through the
Venturi. The moment the tendency to a depression tends
to depress the water-column in the suction tube still
further, a true sucking action commences. So long as the
Venturi end of the suction pipe acts as a piezometer, it is
necessary that this pipe be connected with the outer air
by means of the pet-cock above spoken of, to have its
indications reliable; as otherwise the air contained in the
pipe between the Venturi and the tub of water, or mercury,
at the lower end of the other leg of the suction pipe, may
become compressed or rarified by the action of the water
backing up, or of the Venturi exhausting air, and thus
cause the piezometric readings to be in error. As soon
as a true sucking action commences, however, the pet-
cock must of course be closed.   *   *

The complete records of the experiments are too vol-
uminous to be reproduced in print. Before giving the
results in the tabular, digested form, therefore, some of the
characteristics of the observations taken will be described.

The upper head-gauge, column 5 of the table, was
liable to have large bubbles of air rise up through it when
the water in the forebay fell too low, or was too much
churned up by its discharge through the head-gate. When
this occurred, the experiments were, as a rule, suspended,
and the trouble was remedied by raising the water level
in the forebay, or by causing the water to flow through
longer channels before reaching the cycloidal mouth-
piece. The water in the forebay could not be kept in
sight, and during some of the experiments with a low
stage of water (70-76) an eddy may have formed above
and next the mouth-piece, and carried air into and through
the meter. I regard it probable that the divergence of
experiments (70-76) from the mean was due to this
cause. But during the experiments the indications of
head-gauge No. 1 as to air-bubbles were regarded as con-
clusive, with respect to the presence of air, or of as little

air as was practically attainable, in the water carried by the Venturi meter. When air-bubbles were seen in this gauge-tube, the flow of water to the mouth-piece was ameliorated; and when none were seen, it was supposed that there was no need for the amelioration referred to. At first, readings were taken on this and on head-gauge No. 2, every minute; but after the eleventh experiment they were taken at least every half-minute, and sometimes as many as four per minute, during the duration of the experiment.

The oscillations of the water column in head-guage No. 1 were not large, being in some experiments as little as 0.02 in the course of the experiment, and seldom, if ever, touching 0.10 as their extreme range.

The down-stream head-gauge, column 6 of the table, was more troublesome as to air-bubbles coming up in it, and in its range of oscillations. Air-bubbles could probably have been made to pass by unnoticed, by tapping the piezometer into the bottom of the 1-foot tube; but the original way of tapping it into the top of the tube was adhered to, for the very purpose of being thereby able to judge, somewhat, of the amount of air carried through the meter. Low velocities caused least oscillations, from .05 to .10; high velocities caused greater oscillations in the piezometric column, ranging as high as 0.30, in some cases 0.35, during the duration of a single experiment, but without any effort being made to register maxima and minima. The amount of air carried through the meter was naturally greatest during the experiments with high velocities. It was also governed, no doubt, to some extent by the degree of submergence of the whole apparatus. It was never allowed to be great enough to cast palpable discredit on the indications of the gauge No. 2, having regard to such indications consisting of the average of the many readings (sometimes forty or more) noted down during the course of a single experiment.

The vacuum gauge, column 9 of the table, was supplied with water of very nearly the same temperature as that of the water passing the meter. The gauge alongside of the water-barometer tube was movable, and had for its zero the point of a hook-gauge dipping into the tub of water that supplied the water for the barometric column. The point of this hook-gauge, or zero of the gauge, could thus be constantly kept at the level of the water in the tub.

During extremes of velocity, the oscillations of the water column were least; for velocities of 15. to 35. feet per second through the Venturi they were greater. Two methods of reading this gauge were used. In the one, the observer made his record once a minute, mentally noting, before writing down the observation, what was the average of the oscillations seen. This method gave an extreme range of .10 or .11 during the course of any single experiment. In the other method, the observer noted down heights seen, as fast as he could write, so as to catch the very extremes of all the oscillations. This method, gave sometimes as much as two feet of oscillation during the course of a single experiment. The use of a mercury column instead of a water column would naturally have limited these ranges of oscillation to about $\frac{1}{14}$ of their value as found. But it was deemed best to use the water column for purposes of an accurate representation of the forces at work in the apparatus. On the other hand, the air-chamber was intended to average or to quiet the action of these forces, as they might act through a single orifice bored into the venturi, and directly connected with a piezometric tube or with the suction pipe of a vacuum barometer. At date of this writing, I regard the use of some form of air-chamber as essential to the good working of a Venturi Meter.

When the vacuum column was broken by opening either of the pet-cocks which were above the water level in the air-chamber, the water column, previously supported by the existence of the partial vacuum, would drop instantly, then perhaps oscillate, with a downward tendency in the oscillations.

Its fall could, as a rule, be as instantly arrested, by closing the pet-cock orifice with the finger; and sometimes a number of taps with the finger would be telegraphically repeated by apparently synchronous movements of the water column.

Before presenting the tabular results of the experiments, I also present some remarks as to the methods of computation which yielded these results. The first six columns of the table contain data, and with what has been said above, will need no further explanation.

Column 7 is a mere subtraction of Column 6 from Column 5.

Column 8 will be as readily understood.

Column 9 contains data, being the length of the water column held up in the vacuum gauge, and measured as already described.

Column 10 shows the working of the 3 pet-cocks that were tapped into the air-chamber, one on top, one on line with a horizontal diameter, and one directly at the lowest point of the air-chamber.

Column 11 is the "head on the venturi," or H_v, being the difference of level between the water at, and above the venturi, as indicated by the head-gauge No. 1 and by the vacuum gauge. The elevation of the top of the inside of the venturi was 84.704, and all measured vacuum heights must be subtracted from this quantity to get the constructive elevation of the water, or locus of the hydraulic gradient, at the venturi. The difference between this locus, or elevation of the hydraulic gradient, at head-gauge No. 1 and at the venturi is the "head on the venturi," in the computations as made and recorded in the table.

Column 12 contains the co-efficient belonging to the ordinary computation of discharge through an orifice, when the discharge is as found during the experiment, the orifice is the venturi, and the head is H_v.

Column 13 is the locus of a point at the venturi in a straight line, connecting the points in the hydraulic gradient found at head-gauges Nos. 1 and 2.

Column 14 is the difference between Column 13 and 84.704, plus the measured vacuum heights. It indicates how much the hydraulic gradient was depressed at the venturi below the point given in Column 13. From Columns 13, 14 and 5 was found the H_v to be used in experiments (56-60). With a more perfect experimental apparatus, this and other roundabout methods of interpolation used in computing the tabular results could, of course, have been avoided. But with the data at hand, it was deemed better to utilize all there were, in the best way possible rather than reject any.

Column 15 indicates the characters used in plotting the experiments.

Column 16 is H_v plus the height due the velocity of approach in the one-foot tube; and is called H_v'.

Column 17 is the co-efficient corresponding to H_v' when used similarly to H_v of Column 11.

A star affixed to a number in the table indicates some form of interpolation.

Experiment No. 66 is worthless on account of inability to keep the head steady in the forebay. Experiments 14 and 15 are unreliable on account of too much air in the down-stream head-gauge and passing the meter. And I hold experiments 70-76 to be unreliable, as already stated.

It is much to be desired that the whole series be repeated with the more perfect means for observation, which the experience of this, the first series, suggested.

TABLE No. 1. HERSCHEL ON VENTURI WATER METER.

1	2	3	4	5	6	7	8	9	10	11	12	13	14	15			16	17
No. of experiment.	Duration of experiments, Seconds.	Total volume of water passed. Cu. ft.	Velocity through venturi. Feet per second.	Head Gauges. Upper.	Head Gauges. Lower.	H = Total head. Feet.	Co-efficient, Velocity, $\sqrt{2gH}$	Vacuum gauge, Height of water column. Feet.	Working of pet cocks.	H_v Head on venturi. Feet.	Co-efficient, Velocity, $=\dfrac{H_v}{\sqrt{2g}}$	Hydraulic gradient at venturi.	Depression below hydraulic gradient at venturi.	H_v Curve, a.	Depression curve, b	H_v curve, c	Theoretical H_v.	Corresponding co-efficient.
1	981.25	4226.0	49.70	99.069	88.351	10.718	1.823	24.599	Assume air in all.	38.874	0.994	96.362		O		O	39.355	0.988
2	532.75	2282.4	49.62	99.376	88.320	11.056	1.861	24.512	"	39.184	0.988	96.383					39.604	0.982
3	952.50	4102.4	49.88	99.402	88.365	11.037	1.872	24.531	"	39.229	0.993	96.614					39.714	0.987
4	735.00	3332.0	49.88	99.393	88.359	11.034	1.869	24.599	"	39.288	0.991	96.606					40.771	0.995
5	685.75	3703.3	45.86	94.173	88.390	5.873	2.359	23.685	"	33.154	0.993	92.690					33.564	0.987
6	483.75	2917.1	45.90	94.184	88.340	5.844	2.367	23.655	"	33.135	0.994	92.708					33.545	0.988
7	587.50	2608.3	43.53	94.254	89.400	4.854	2.464	19.704	"	29.254	1.005	93.028					29.623	0.997
8	593.75	2221.2	43.33	94.238	89.560	4.658	2.456	10.339	"	28.873	1.007	93.388					29.239	0.999
9	600.00	3057.7	39.72	94.344	90.562	3.784	2.546	14.965	"	24.205	0.986	93.380					24.573	1.000
10	579.00	2048.4	38.98	94.335	90.557	3.778	2.593	14.672	"	24.300	0.995	93.381					24.595	0.980
11	231.00	846.7	39.07	94.437	91.557	3.778	2.506	14.348	"	23.979	0.999	93.724					24.276	0.949
12	504.75	2461.9	38.85	94.417	91.643	2.784	2.522	8.251	"	17.974	0.994	93.906					18.196	0.986
13	705.75	2242.7	33.85	94.417	91.603	2.814	2.516	8.307	"	18.080	0.943	93.104			O	O	18.243	0.968
14	565.50	1393.2	28.43	91.599	89.610	1.990	2.508	7.215	"	14.150	0.943	91.104					14.777	0.938
15	469.25	1559.7	28.52	91.599	89.585	2.014	2.515	5.704	"	12.599	1.005	91.099					12.758	1.002
16	494.75	1595.3	28.15	91.605	89.649	1.956	2.520	5.787	"	11.980	1.004	91.111					12.234	1.008
32	351.50	5559.9	52.10	99.203	88.367	10.836	1.936	24.828	"	39.327	1.026	96.466					39.896	1.010
33	378.50	5650.5	50.51	99.155	88.378	10.777	1.918	24.681	"	39.132	1.007	96.433					39.602	1.000
34	493.25	5752.4	50.33	99.227	91.023	8.190	2.192	24.115	"	38.634	1.000	97.151					39.127	1.003
35	424.75	5715.6	50.38	99.240	93.071	5.509	2.472	19.358	"	33.894	0.002	97.833					34.300	0.996
36	426.75	1469.6	39.84	99.444	95.665	3.779	2.558	10.018	"	24.758	0.997	98.489	9.612	O	O	O	25.667	0.993
37	375.75	818.8	35.24	94.200	92.600	1.600	2.488	0.520	Top draws air. T. & M. air. B. Water.	10.016	0.994	93.796			O	O	10.140	0.988
38	535.00	986.8	21.36	91.650	90.500	1.150	2.484	0.375	All air.	7.321	0.984	91.360	7.031				7.410	0.979
39	481.75	936.5	22.28	92.350	91.100	1.250	2.485	0.556	"	8.200	0.970	92.034	7.886				8.208	0.965
40	518.75	1177.9	26.30	91.150	93.498	1.652	2.551	0.983	"	11.009	0.988	94.733	10.592				11.143	0.982
41	592.25	1196.5	27.99	95.150	93.350	1.800	2.564	1.643	"	12.089	0.989	94.695	11.634				12.237	0.923

No.									Description						
43	441.00	1098.3	28.84	95.500	93.750	2.000	2.543	2.511	All air.	13.307	0.986	94.995	12.802	13.469	0.986
43	428.74	1102.3	29.78	95.454	93.350	2.104	2.569	3.463	"	14.213	0.985	94.923	13.682	14.385	0.972
44	312.00	885.6	32.88	95.380	92.900	2.520	2.582	5.520	"	17.002	0.994	94.783	16.965	17.212	0.988
45	361.25	1109.2	35.56	95.380	92.350	3.030	2.547	9.342	"	20.018	0.991	94.615	19.253	20.264	0.985
46	143.50	512.8	41.39	95.600	91.397	3.993	2.583	15.112	"	25.728	1.018	94.291	24.699	26.041	1.011
47	231.25	940.0	47.08	95.600	90.000	5.600	2.581	24.045	"	34.941	0.993	94.186	33.527	35.373	0.987
48	252.25	1071.4	49.19	95.600	88.200	7.490	2.755	25.477	"	36.373	1.042	93.731	34.504	36.844	1.011
49	409.50	1158.9	32.78	95.443	93.600	2.443	2.615	6.097	"	16.836	0.996	94.826	16.219	17.045	0.990
50	433.00	1091.4	29.19	95.500	93.600	1.900	2.641	2.299	" .	13.035	1.008	93.020	12.555	13.201	1.002
51	252.75	606.6	27.80	95.500	93.750	1.750	2.620	1.038	T. air.	11.834	1.008	95.958	11.392	11.984	1.001
52	464.00	1052.8	26.28	95.900	93.900	1.600	2.590	0.246	M. air & water. B. water.	11.042	0.986	95.096	10.638	11.176	0.980
53	669.00	1051.9	18.21	88.896	88.053	0.843	2.473	1.171	All air. T. & M. air.	5.363	0.980	88.683	5.150	5.428	0.975
54	557.50	814.2	16.92	88.922	88.191	0.731	2.467	0.394	B. water. T. air.	4.612	0.982	88.737	4.477	4.667	0.976
55	625.50	904.7	16.75	88.912	88.219	0.693	2.599	0.211	M. & B. water.	4.419	0.994	88.737	4.344	4.474	0.988
56	595.25	616.7	14.14	88.999	88.439	0.500	2.493		Water level at the venturi above the air-chamber.	3.20	0.985	88.833	3.07*	3.24	0.977
57	540.00	398.9	8.56	88.960	88.766	0.200	2.385			1.21	0.970	88.999	1.16*	1.32	0.966
58	574.75	217.5	4.38	88.940	88.890	0.030	2.444			0.36	0.911	88.97	0.35*	0.36	0.911
59	754.50	87.6	1.35	88.971	88.961	0.010	1.677			0.05	0.750	88.968	0.05*	0.05	0.750
60	748.25	144.0	2.23	88.971	88.951	0.019	2.016			0.10	0.879	88.966	0.10*	0.10	0.879
61			18.00*	88.850	88.050	0.800		1.005	All air. T. & M. air. B. water.	5.151	0.984	88.648	4.949	5.214	0.983
62			17.60*	88.850	88.086	0.770		0.954		5.100	0.972	88.655	4.905	5.160	0.966
63			17.12*	88.890	88.120	0.730		0.563	T. water. M. & B. water.	4.709	0.984	88.665	4.524	4.766	0.998
64			19.00*	88.880	88.170	0.690		0.283		4.439	0.982	88.685	4.264	4.493	0.976
65			16.05*	88.865	88.220	0.645		0.019	T. air. M. & B. water.	4.186	0.979	88.702	4.017	4.730	0.973
66			12.00*	88.350	87.976	0.374		0.008	All air.	3.654	0.783	88.255	3.559	3.68	0.780
67			15.50*	88.016	88.010	0.606		0.234	"	4.156	0.948	88.463	4.003	4.203	0.943
68			19.83*	89.030	88.070	0.862		1.268	"	6.094	1.002	88.787	5.851	6.171	0.995
69			18.72*	88.915	88.053	0.906		1.332	"	5.543	0.991	88.647	5.335	5.612	0.985
70			14.10*	86.806	85.767	0.659		2.084	"	2.936	1.026	86.678	2.838	2.975	1.019
71			16.20*	86.426	86.039	0.393		2.084	"	3.866	1.035	86.259	3.639	3.857	1.029
72			12.30*	86.432		0.534			"	2.262	1.000	86.333	2.163	2.292	1.043
73			11.25*	86.484	86.152	0.332		0.160	T. air. M. & B. air.	1.940	1.007	86.400	1.856	1.965	1.001
74			10.40*	86.399	86.117	0.282		0.015	" T. & M. air.	1.710	0.992	86.328	1.539	1.731	0.986
75			8.35*	85.892	85.699	0.193		0.104	B. water.	1.292	0.916	85.843	1.243	1.305	0.911
76			9.73*	86.006	85.754	0.252		0.227		1.529	0.981	85.942	1.465	1.547	0.975
77			22.38*	94.28	93.08	1.20		0.00							

The October set of experiments were made with a Venturi Meter built up inside of the 9-foot trunk that feeds the testing flume. This trunk is of the sort usually built in Holyoke, of boiler iron, to serve as a penstock of mills.

Every alternate ring, or section of about $4\frac{1}{2}$ feet in length, is a spigot-piece, at both ends, and the others are bell-pieces. Each ring is formed of three plates, lapped and riveted, the bell and spigot-joints also being riveted. Nominally 9 feet in diameter, very careful measurement on 15 diameters, at the 35 narrowest sections of a length of about 150 feet, gave the average area, rivet-heads subtracted, 57.823 square feet. The area of the "average shape" was 57.742 square feet.

This "average shape" was very nearly a true ellipse, having 8.70 and 8.93 for its minor and major axes, the minor axis being the vertical one.

Commencing at the upper level canal, comes the usual rack to keep out floating substances; then a head-gate, in these experiments wide open, and of no influence; then a piece of trunk, about 7 feet in diameter and 22 feet long; then a conical piece, 8 feet long, to expand from the 7-foot to the 9-foot trunk; then the 9-foot trunk, in a straight line with the two pieces already named, 224 feet, then curving on a quarter circle of 40 feet radius of center line, and a straight length 9 feet long, to the vestibule of the testing flume, as indicated in Plate XXXV.

Head-gauge No. 1 was situated 73.92 feet down-stream from the inside of the rack, the inside of the rack being 35.5 feet up-stream from the up-stream end of the 9-foot trunk. From head-gauge No. 1 to the beginning of the up-stream cone was 36.81 feet; thence to center of the venturi, 16.10 feet; thence to end of lower cone, 69.58 feet; thence to head-gauge No. 2, 30.39 feet.

This trunk has an inclination down-stream of 1.577 in 100 feet, as measured from the interior surfaces of the orifices leading to head-gauges Nos. 1 and 2. All the structures placed in it, hereafter to be described, were set in planes at right angles to the center line, or so as to have the center line for their geometric axis. The inside of the top of the venturi was on grade 90.909. The water-level in the upper canal at this point is about 99.90; that of the lower-level canal about 79.90.

As above stated, the general dimensions of the meter were intended as given in Plate XXXIII. The venturi is shown in Plate XXXVII, and was made, as in the 1-foot meter, of cast-iron, lined with brass; differing from the 1-foot venturi, however, in having eight separate air-chambers, one for each ¼-inch hole leading out of the venturi.

These several air-chambers had each a suction-pipe attached, and a pet-cock, as shown in Plate XXXVII. The several suction-pipes had each a stop-valve, and were then assembled into the main suction-pipe, which led to the vacuum-gauge or water-barometer. By means of this arrangement, any one, all, or a combination of several air-chambers could be connected with the vacuum gauge to the exclusion of the others.

Plate XXXVIII shows the whole meter in longitudinal section. The iron and brass venturi was, in this case, as will be seen, a true cylinder, about 3 feet in diameter and only 1 foot long. Its area was 7.07425 square feet, as determined from a measurement, with a vernier slide-rod, on 12 diameters; 4 at each end and at the center of the cylinder.

At either end of the venturi was a wooden connecting-piece, made to flare outwardly in the curves shown in Plate XXXIII, and built up of staves, hooped with strong cast-iron frames, turned and smoothed in a lathe and soaked in water before setting. At either end of the central portion thus formed came the two cones, consisting of planed pine strips nailed to circular frames or hoops, set inside the 9-foot trunk. This construction of the cones did not leave them so smooth, at all the joints and butts, as was the case with the 1-foot cones, of course; the whole surface was, however, much smoother than the interior of the iron trunk at either end. Two water-tight bulk-heads set in the trunk at either end of the central portion above referred to, and a man-hole cut into the trunk from the outside between these bulk-heads, gave access to the outside of the venturi during the experiments.

In these experiments the head-gauges were hook-gauges, measuring water-levels inside of stout boxes, that were connected with the 9-foot trunk by $\frac{3}{4}$-inch pipes. These $\frac{3}{4}$-inch pipes were connected with short brass ajutages let into the shell of the trunk, and smoothly filed off on the inside. The trunk itself was tapped to receive these ajutages, in the case of each gauge, at the top of one of the smaller or spigot rings, about 1 foot up-stream from its entrance into one of the larger or bell rings.

The quantity passing the meter was measured over the weir of the testing flume. This is a permanent weir, having a wrought-iron sharp edge, which can be used without end-contractions on a length of about 20 feet, or with end-contractions on shorter lengths.

It was used with end-contractions on a 6-foot length in experiments (1-7), and without end-contractions in the remaining experiments. About 10 feet up-stream from the weir is a horizontal perforated brass tube set some 9 inches above the floor (which in turn is 5.9 feet below the crest of the weir), this tube being connected with a galvanized iron bucket set in a recess in the wall and fitted with a hook-gauge for measuring depths upon the weir. I have always used a light leveling-rod, $\frac{3}{4}$-inch square, graduated on all four sides, by Darling, Brown & Sharpe, of Providence, R. I., brass-tipped at the ends and set directly on the hook, to compare weir heights with the

PLATE XXXVII.
TRANS. AM. SOC. CIV. ENG'RS.
VOL. XVII NO. 371.
HERSCHEL ON
VENTURI METER.

Venturi Experiments.
Scale of feet.
8
7
6
1
2
3
4
5

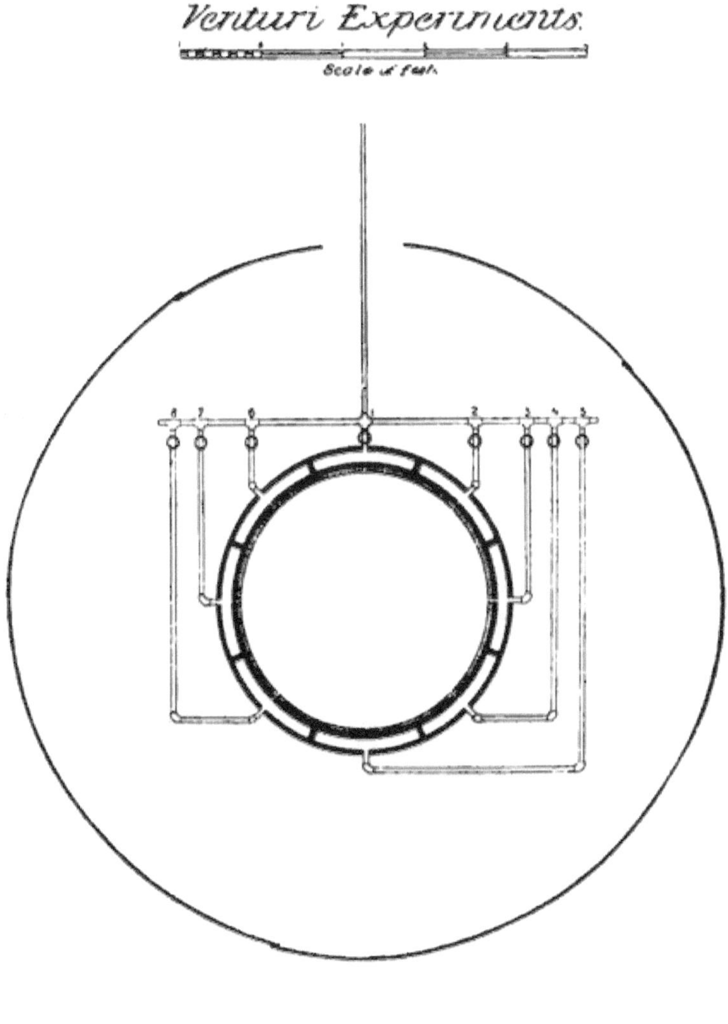

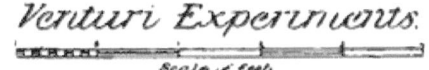

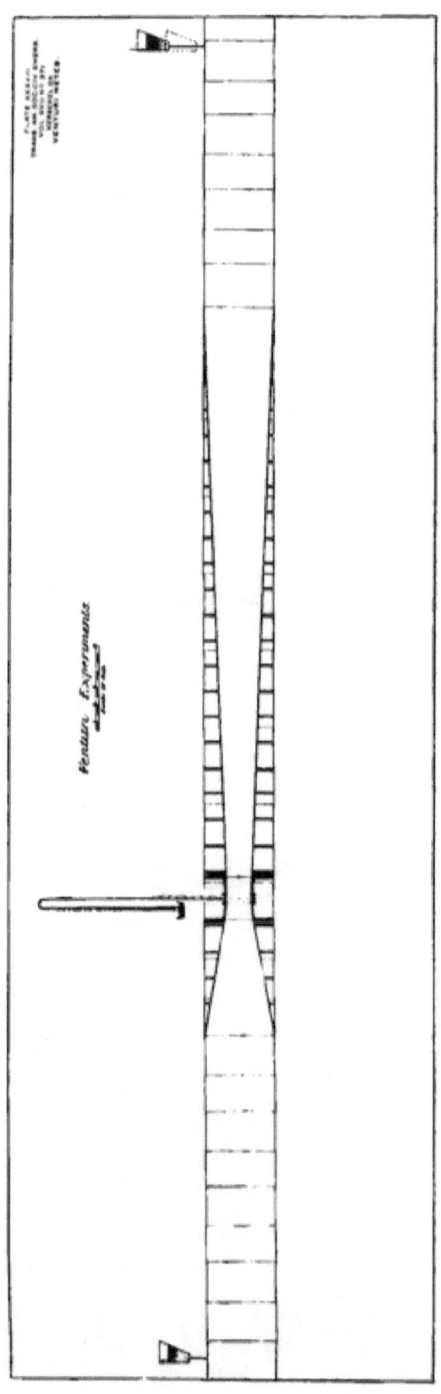

setting of the hook-gauge. Two racks deaden the water before it reaches the line of brass tube that leads to the hook-gauge bucket.

As some of the weir heights exceed the limit of the Francis experiments, I used in the computations of these quantities the co-efficients given in Hamilton Smith's "Hydraulics."

These are the result of a careful sifting and digest of all attainable original publications of the records of reliable experiments on the discharge over weirs, inclusive of those made by James B. Francis, and recorded in "Lowell Hydraulic Experiments." Their results, differ from the results of the Francis formula in the present instance as follows, giving a few characteristic differences.

Depth on the Weir.	Difference to Reduce F. Quantity to S. Quantity.	
	6-foot Weir.	20-foot Weir.
	Per cent.	Per cent.
0.2	+ 1.7	+ 2.0
0.8	+ .0	− 0.7
1.0	+ .1	− 0.45
2.0	+ 1.0	+ 1.0
2.5	+ 1.4	+ 1.5

The measurement of leakages was conducted with the same care that obtained in the June experiments already spoken of. They could in all cases be measured by the variations in the water-level of a defined area, the proper allowance being then made for an increased or a diminished head upon the orifices causing leakage for the duration of each experiment; in one case, that of the two bulk-heads either side of the venturi, the leakage, an insignificant quantity, was pumped out and measured in pails. The total leakage never exceeded 0.72 cubic feet per second, and ranged from that down to 0.47, or from

about 0.3 to 4 per cent. respectively, of the quantity passing the meter.

In these experiments, also, the discharge of the meter
was a submerged one. The water at the extreme down-
stream end of the 9-foot trunk always stood higher than
the top of the trunk.

Before setting the Venturi meter into the 9-foot trunk,
but after the head-gauges subsequently used in the Venturi
meter experiments had been established, a series of experiments were made to test the loss of head in the original
9-foot trunk. The results of this series are given in the
table which follows.

The area was taken = 57.823 square feet as above
stated.

$D = 8.58$, though the "average shape" was an ellipse
and not a circle.

$l = 152.88$, being the distance between centres of
piezometric brass ajutages above spoken of.

Each experiment is based upon the average result of
not less than forty consecutive half-minute readings.

In this and the following series of experiments, taking
their supply of water from the upper level canal at
Holyoke, an extra gate tender was stationed on duty to
keep the upper level steady. By constantly wasting out
of the upper level, at a point some distance from the
head-gates, then keeping the water steady at this point
by regulating the amount wasted, and with the aid of a
canal a mile long and averaging 125 feet wide, as was the
case, this was reasonably well accomplished. During the
course of a single experiment the canal seldom varied as
much as 0.10 in level. Then, in the well-known formula
$v = n\sqrt{rs}$, s being equal to $\dfrac{h}{l}$ (see Hamilton Smith's "Hydraulics," p. 271,)

v, h and n have the following corresponding values.

v	h	n
Feet per second.	Feet.	Co-efficient.
0.5	0.0012	121.9
1.0	.0049	120.6
1.5	.0128	111.9
2.0	.0238	109.4
2.5	.0375	109.0
3.0	.0548	108.2
3.5	.0763	107.0
4.0	.1012	106.2
4.5	.1295	105.6

The results of the thirteen experiments made, plotted so regularly, and the points were so close together, that there is hardly a choice in reliability between points taken from plotted curves at regular intervals, as given in the table, and the direct results of experiment. I judge from the disagreement of the results above given, with those found at other places, but on longer tubes, either that piezometers do not correctly indicate the h of the formula (see Hamilton Smith's "Hydraulics,") or else that a uniform and non-accelerative régime of the flow of water through the trunk had not become established in the comparatively short length at command for purposes of measurement. This latter circumstance would, in most cases, prevent any attempt to compute the flow of water through mill trunks, and in many cases of city water pipes, by the use of the formula $v = n\sqrt{rs}$, or of any other formula for the discharge of pipes; whose general co-efficients can only be established for the case of a perfectly uniform, permanent flow; three modifying conditions, namely, those of perfection, uniformity, and permanency, which are very difficult to obtain in practice.

Passing now to the description of the table about to be given:

Columns 1 to 8 will hardly need explanation other than that already above written. All the data tabulated are the averages of consecutive half-minute readings, except in case of the head on the weir, where the usual one-minute interval between readings was adhered to.

Columns 9 and 10 will be clear, when it is remembered what was above said with regard to depression of the piezometric water column in the branch of the suction-pipe which leads directly out of the air-chamber, and concerning the vacuum-gauge formed by the other, descending leg of the same suction-pipe; remembering, moreover, that the plane of division between the action of these two forms of the venturi gauge lies on grade 90.909, as above stated.

Column 11 brings us to a peculiarity in this method of metering water, first revealed by this series of experiments, viz., that the indications of the venturi depression, or vacuum gauge, are different, according as the venturi has been pierced upon a different diameter.   *   *

As these experiments were made for the sole purpose either of discovering or of perfecting a new and practical method of gauging water, I have not pursued the study of this apparent idiosyncrasy of the meter any further than as stated in the Table. As I translate the results, they mean that the venturi must, in all cases, be pierced for connection with the air-chamber vertically at its crown, and may be pierced radially at as many additional points as we please, without affecting the reading of the standard crown orifice. I have no experimental results to guide me in a choice between several venturi orifices and air-chambers (one at the crown always included), or only a crown orifice; or between air-chambers separate and distinct for each orifice, as in the October experiments, and an air-chamber common to several orifices and encircling the venturi, as in the June experiments. Still, as my feeling on the subject, being that of a person who has worked with the two forms of meter, may be interesting, I will state that, at present, I should favor the general form of air-chamber and of orifices used in the June experiments.

Column 12 will need no explanation. In this set of experiments, with the glass suction-pipe next the venturi and the separate air-chambers, these pet-cocks were of subordinate value. To open a pet-cock, so long as the depression-gauge (not vacuum-gauge) is acting, does not disturb the piezometric water column.

Columns 13 to 16 demand no further explanation than was given for the similar columns in the table relating to the June experiments.

As regards the range or oscillations of the several gauges, in the space of one experiment, during this series, it was:

```
In head-gauge No. 1, seldom so much as............0.05 feet,
   "     "     " 2,     "       "     ............0.05  "
In the depression-gauge,  "      "     ............0.11  "
In the vacuum-gauge,      "      "     ......0.30 to 0.50  "
```

as a result of ½-minute readings, and with no attempt to record absolute maxima and minima.

The quantity of water discharged must have been very nearly uniform in its flow per second.

I pass now to a brief discussion of the results as found in Tables No. 1 and 2.

Consideration of Results Shown by Tables.

As a measure of comparison of the uniformity of the results found with the two Venturi Meters experimented on in June and October, 1887, I suggest the range of co-efficients known to exist in the case of the weir, or of a simple orifice. The weir has hitherto been regarded as a standard method of gauging water; yet every one who has practiced with it knows how carefully all its dimensions must be proportioned, and the water led to it, in

TABLE No. 2.
HERSCHEL ON VENTURI WATER METER.
VENTURI EXPERIMENTS.

Observations were taken every half minute; 16 minutes' "Duration," therefore equal 33 observations, and similarly in all the experiments.

1 No. of experiment.	2 Duration of experiment in minutes.	3 Quantity of water passing venturi. Cub. ft. per sec.	4 Velocity through the venturi. Feet per second.	5 Head gauges. Upper. Feet.	6 Head gauges. Lower. Feet.	7 H = Total head. Feet.	8 Co-eff.$=\dfrac{\text{Velocity}}{\sqrt{2gH}}$	9 Depression gauge. Feet.	10 Vacuum gauge— Height of water column. Feet.	11 Valves. Nos. open.	12 Pet cocks.	13 H_v = Head on venturi. Feet.	14 Co-eff.$=\dfrac{\text{Velocity}}{\sqrt{2gH_v}}$	15 Theoretical $H_v=$ Feet.	16 Theoretical co-efficient.
1	16·	12·40	1·75	99·8830	99·8983	0·0047	3·188	99·8441	.....	1	.....	0·0389	1·108	0·0397	1·097
2	10·	12·36	1·75	·8765	·8714	·0051	3·051	·8480	.....	5	.....	·0276	1·311	·0276	·991
3	10·	12·40	1·75	·8930	·8875	·0055	2·947	·8538	.....	3	.....	·0392	1·104	·0392	·987
4	20·	35·19	4·96	·8820	·8374	·0446	2·929	·4982	.....	2	.....	·3639	·998	·3897	·986
5	20·	54·02	7·64	·8383	·7361	·1022	2·978	98·9209	.....	All	.....	·9174	·994	·9311	·974
6	20·	53·98	7·63	·8433	·7415	·1018	2·968	·8803	.....	5	.....	·9165	·994	·9302	·972
7	10·	53·91	7·62	·7966	·6948	·1018	2·978		.....	5	.....	·9163	·993		·973
8	20·	71·11	10·05	·8403	·6652	·1751	2·995		.....	All	.....	1·6286	·982	1·6496	·974
9	9·5	70·99	10·03	·8333	·6584	·1749	2·992		.....	1	.....	1·6261	·981	1·6467	·972
10	9·	79·80	10·01	·8250	·6575	·1725	2·993		.....	3	.....	1·6035	·929		·973
11	20·	88·76	12·55	·7731	·4993	·2733	2·993	97·2222	.....	All	.....	2·5509	·980	2·5858	·974
12	20·	103·11	14·58	·7090	·3692	·3600	3·029	96·2273	.....	..	.....	3·4317	·981	3·4815	·974
13	20·	119·96	16·96	·6358	·1400	·4958	3·003	94·9873	.....	..	.....	4·6485	·981	4·7158	·968
14	19·	136·15	19·67	·5278	98·8574	·6704	2·995	93·1935	.....	..	.....	6·3343	·974	6·4246	·961
15	20·	162·40	22·96	·4179	·5175	·9054	3·009		0·2495	5	.....	8·7584	·968	8·8820	·952
16	8·5	162·60	22·98	·4305	·5246	·9063	3·010		·0042	1	.....	8·5261	·981		·952
17	20·	162·60	22·98	·4124	·5071	·9053	3·012		·2516	..	.....	8·7550	·969		
18	9·	162·60	22·98	·4094	·5037	·9057	3·011		·0011	2 and 6	.....	8·5015	·983	8·8786	
19	4·5	162·60	22·98	·4129	·5129	·9000	3·021		·2400	1, 2 and 6	.....	8·7439	·969	8·8675	

20	4·5	162.600	22.98	.4033	.5045	.018	3.018	……	.2300	1	No. 1 air	8.7243	.970	8.8479	.963
21	20.	183.99	16.00	.2618	.1097	1.1521	3.020	……	2.7156	1	No. 2 water	11.0684	.974	11.2266	.967
22	9·	183.62	25.96	.2710	.1172	1.1537	3.013	……	2.6979	1, 2 and 6, 3 and 7	1, 2 and 6, 3 and 7 air	11.0599	.973	11.2271	.966
23	9·5	183.62	25.96	.2391	.0996	1.1395	3.032	……	2.3189		Nos. 4 and 8, 5 water	10.6490	.991		
24	4·	183.99	26.00	.2618	.1204	1.1414	3.034	……	2.6689	1, 2 and 6, 3 and 7		11.0217	.976	11.1799	.969
25	4·5	18.48	25.94	.2447	.0976	1.1471	3.019	……	2.7160	1		11.0517	.973	11.2089	.966
26	20·5	207.24	29.30	.1200	97.6337	1.4863	2.996	……	6.1874	All	All air	14.3984	.963	14.5988	.956
27	19·	207.39	29.37	.1391	.6466	1.4925	2.992	……	6.1982	1		14.4283	9.62	14.6298	.956
28	20·6	207.09	29.27	.1330	.6475	1.4855	2.995	……	5.9954	5	No. 5 air	14.1754	.964		
29	10·5	206.79	29.23	.1174	.6797	1.4877	2.988	……	6.0086	4	No. 4 air	14.3070	.964		
30	4·	206.95	29.25	.1089	.6497	1.4729	2.998	……	6.1389	4 and 8	No. 8 air	14.3388	.965		
31	4·5	206.64	29.21	.1052	.6497	1.4820	2.988	……	6.0505	3 and 7	Nos. 3 and 7 air	14.2467	.964		
32	5·5	206.64	29.21	.1015	.6688	1.4750	2.999	……	5.8680	2 and 6	Nos. 2 and 6 air	13.9752	.965		
33	2·	206.95	29.25	.1062	.6386	1.4774	3.002	……	5.8762	2		14.0052	.974		
34	3·5	207.39	29.25	.1209	.6421	1.4823	3.002	……	6.1668	6		14.4881	.975		
35	8·5	206.95	29.25	.1301	.3933	1.4880	2.990	……	7.6815	All	All air	14.3879	.962	14.5883	.955
36	19·5	217.51	30.75	.0299	.3837	1.6366	2.997	……	7.7983	1		15.8024	.964	16.0234	.958
37	14·5	217.51	30.75	.0095	.4094	1.6658	3.007	……	7.7157	1 and 5	No. 5 air	15.8988	.961	16.1198	.955
38	13·5	217.65	30.77	.0322	.4019	1.6228	3.011	……	7.8409	3 and 7		15.8189	.961		
39	5·	217.65	30.77	.0227	.3971	1.6203	3.014	……	7.5550	4 and 8		15.9541	.960	16.1751	.954
40	3·5	217.51	30.88	.0231	.4315	1.6260	3.006	……	7.6427	All		15.6631	.968		
41	20.	218.45	32.91	.0546	.0956	1.6731	3.022	……	10.2205	1	No. 1 air	18.2364	.969	18.4903	.954
42	10.	232.83	32.98	98.9249	.0829	1.8653	3.005	……	10.2314	3 and 7	Nos. 3 and 7 air	18.2739	.961	18.5285	.955
43	9·	233.31	32.94	.9515	.0774	1.8826	3.008	……	9.9032	4 and 8	Nos. 4 and 8 air	17.9404	.962		
44	16·5	233.00	32.96	.9462	.1002	1.8608	3.004	……	10.0259	All		18.6430	.910		
45	12·	233.16	34.75	.9070	96.8216	1.8668	3.008	……	12.1940	1		17.9404	.966		
46	14·	245.84	34.87	.9134	.7241	2.0493	2.996	……	12.2045	3 and 7	Nos. 3 and 7 air	20.1261	.964	20.4805	.957
47	9·5	243.87	34.43	.7734	.7077	2.0516	3.003	……	12.0516	1		20.0683	.959	20.3471	.953
48	4·	243.54	34.56	.7593	.7536	2.0698	2.993	……	12.8222	5		19.8633	.963	20.3761	.955
49		244.52		.8234			2.996			All		20.0916	.961		

order that it may give truthful or accurate results. The range of the co-efficients entering into any proposed formula for the discharge over a weir, can be and has been limited by limiting the general and proportional dimensions of the apparatus and water depths to which the formula was to be applicable. And the most positive results are undoubtedly found by taking such a limited formula, constructing the weir or other apparatus in accordance with the dictates of the experiments on which it was founded, and then, practically, by repeating the experiments, repeating the attainment of the original results. And wherever this can be done with a weir, the method of experiments made at Lowell by James B. Francis, and the formula based upon them will, no doubt, long remain the standard method and formula for weir measurements. But without such close limitation and imitation, or taking depths upon the weir ranging only from 0.3 to 2.0 feet, and taking weirs both with and without end-contractions, the co-efficient varies from .660 to .580 (see Plate VII, Hamilton Smith's "Hydraulics,") or 12 per cent.

Taking only weirs without end-contractions, the range is from .660 to .614, or about 7 per cent. In case of the 1 foot Venturi Meter, and velocities through the venturi, ranging from 5 to 50 feet per second, this range of co-efficient was 6.5 per cent.; in case of the 9-foot Venturi Meter, and velocities through the venturi, ranging from 5 to 36 feet per second, it was 3.3 per cent. But better than this is the fact that the two meters, though differing so much in their size and structure, showed a total combined range of co-efficient no greater than the smaller one alone, or only 6.5 per cent.; taking the co-efficients based on Hv', being those corrected for velocity of approach, and using that simplest of all hydraulic formula, $v = \sqrt{2g\,Hv}$.

Though the areas of discharge were as 81 to 1, and the interior fractional surfaces were widely different, the resultant co-efficients are at extreme points only 6.5 per cent. apart; and the deviation of any single experiment from the resultant mean is 3.0 per cent. in case of the 1-foot tube, excluding the unreliable interpolated results, and only $\frac{1}{2}$ per cent. in case of the 9-foot trunk for its whole range of velocities.* If we compare this to the case of a discharge through various orifices, the result is still more gratifying. To the wearied sojourner among such tables of discharge — ranging in their co-efficients from the familiar 0.6 or $\frac{3}{5}$, up to the mystical co-efficients in the eighty's and ninety's, said to have been found by some one "on large sluice-gates in France" (and occasionally met with in the current practice of the hydraulic engineer)—a consistency in co-efficients as above found for the hydraulic apparatus herein described, is indeed refreshing. We appear to have here, at last, an apparatus for gauging liquids which may range in its dimensions, its materials of construction and in manner of use, so as to cover all ordinary practice, and yet have only 6.5 per cent. of range of co-efficient, at the same time requiring only the simplest of observations and of formulas to work with. Or by limiting the use of the meter to velocities greater than 9 feet per second through the venturi, being about 1 foot per second through the pipe thereto appurtenant, all the ranges of variations above given become materially less.

I said above that we "appear" to have such an apparatus, or, to completely express the underlying thought, we appear to have it in the light of the only two sets of experiments yet made, so far as I know, with a Venturi Meter set in line of a pipe. But further experiment will be needed to confirm or upset such a conclusion,

---

* See Plate XXXIX, showing plotted results.

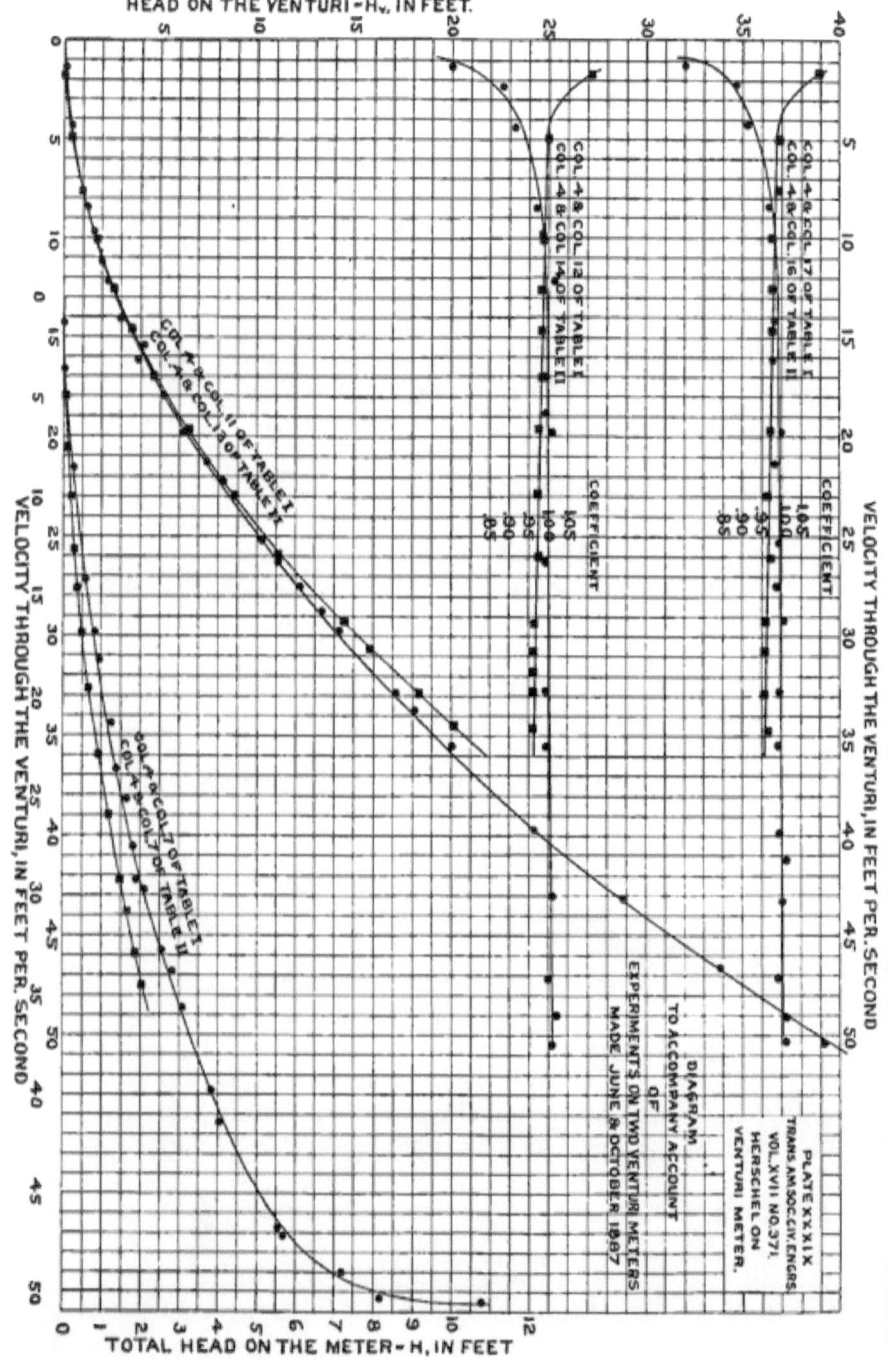

DIAGRAM
TO ACCOMPANY ACCOUNT
OF
EXPERIMENTS ON TWO VENTURI METERS
MADE JUNE 18 OCTOBER. 1887

PLATE XXXIX
TRANS AM.SOC.CIV.ENGRS.
VOL.XVII NO.371.
HERSCHEL ON
VENTURI METER.

and one object of the present paper is to invite such further experiments.*

The reason, I will suggest, why the co-efficients belonging to this form of gauging apparatus are so nearly uniform, is largely on account of the close similarity between the conditions assumed by theory and those found in actual practice, regarding now the state of the liquid as it passes through the venturi. Here, if anywhere, water may be supposed to flow as though composed of the traditional " filaments" of the school-books; while the bubblings of a boiling, seething caldron are but little more violent and irregular than the motions of the alleged "threads" of water, as we find that water in ordinary practice, and as it flows in canals or even in the ordinary line of pipes, or in tubes.

Still the co-efficient is not the same for all velocities; it is less for higher velocities than it is for lower ones in the October experiments, while the reverse holds true in the June experiments; the meter does not appear to be applicable for velocities below 5 feet per second† through the Venturi, or about ½-foot per second through the pipe in which it is placed; and the co-efficient is not equal to one, except in one instance.

The difference between the equation giving the locus of the co-efficients as applicable to the 9-foot trunk and to the 1-foot tube, may be due to difference in asperity of their interior surfaces; some of it may possibly be due to the shortening of the 9-foot cones, caused by the trunk measuring only 8.7 feet high instead of 9 feet as supposed.

---

* While this paper is being written I am in receipt of the October number of the *Journal of the Association of Engineering Societies*, which gives the results of experiments upon similar forms of discharge, but discharging from a tank, and through an orifice of only about 0.03 feet in diameter, and in which the same co-efficient is likewise found nearly equal to 1.

† For the case of $v = 0$, Hv is also $= 0$, and the co-efficient becomes $= \frac{0}{0}$ which may be any assignable quantity.

This justifies the curiously diverging form of the curves of co-efficients for the two Venturi Meters, as shown on the diagram.

Part of the deficiency from the value 1. may be due to defective guidance of the water as it approaches the venturi. It would have been better, no doubt, to have rounded off the angle with which the up-stream end of the smaller cone meets the up-stream pipe or trunk; better still, to have made that portion of the meter up-stream from the venturi of a form which would be generated by the revolution about the central axis of an ogee curve. In the case of the discharge from an open canal or from a tank, this portion of the meter could be suppressed entirely, and in its stead be placed only a mouth-piece, having the form of the contracted vein to feed the venturi; with a head-gauge reading directly the water-level in the tank or canal.

It is also an interesting question whether the vacuum-gauge is indicative of the mean velocity, or of the velocity of the exterior filaments of the body of water passing through the venturi, or of both, and what is the precise meaning of the readings of this and other piezometers tapped into pipes of flowing water. In our present very imperfect knowledge of the action and precise meaning of the indications of such piezometric columns, especially when applied to tubes, but little can be positively affirmed about them.

Loss of head is still the only difficulty to contend against in the practical application of the meter for mill purposes. For purposes of metering a city, or domestic water supply, or water used for purposes other than power in mills, this loss is insignificant. In the other case named, and for a 9-foot trunk, it would be about 1 foot, when the mean velocity through the trunk was 2.7 feet, and $\frac{1}{2}$-foot for a velocity of about 1.9 feet. If the circumstances are such that this loss of head is not permissible, or cannot be paid for by the delivery of enough more water to yield to the consumer an equivalent amount of power, then this meter cannot be used in a form that would make it continuously the sole outlet or inlet of the water to be metered. It could be applied in those cases either at the inlet or outlet, in the penstock or in the tail-race, but would have to be provided with some form of byepass to be kept open at all those times when the operation of metering the water was not actually going on. This could be readily done in the case of an open feeder or of most any tail-race, and as the operation of metering need require so little time, barely five or ten minutes, there could hardly be any objection made by the consumer to this plan of measuring water. It probably need not be pointed out that the whole apparatus could literally be submerged, or covered with water, and yet be conveniently used and act as it ought to, so long as it afforded the only outlet from one body of water to another, and that its advantages in freedom from any moving parts, and from liability to be stopped up or put out of order by floating substance or by ice, are very great.

Writing so soon, only a few weeks after the close of the second set of experiments, I do not very likely allude to all the capabilities of the meter, and have hardly broached the interesting subject of the theory of the instrument. It seems to me that it may in many instances replace the use of a weir, being easier applied and equally or more accurate, and it can be used where a weir is entirely inapplicable.

Mr. CLEMENS HERSCHEL (in reply to questions asked at the time of reading the paper).—In its completed form, the Venturi Meter is an instrument to gauge the quantity of water flowing in a pipe, by measurement of an abrupt, artificially produced depression in the hydraulic gradient. To explain more particularly, suppose a pipe full of water the water in a state of rest. Piezometers placed on such

Discussion of Paper.

a pipe, will have the water stand in them at points situated all on one level.

In Plate XL, suppose PP to be such a pipe, at one time "submerged," or under a head, to the extent 1 P, and again, only to the extent a P. Next, suppose the water to take any velocity through the pipe (no meter being yet supposed inserted), sufficient to cause the water in the piezometer to stand on the line 22 and bb, respectively, according as the amount of submergence was originally 1 P or a P. This line 22, or bb, is what I have called the "hydraulic gradient." Next, suppose the meter inserted in the pipe; upon which, the water level at the up-stream piezometer will remain at 2, but at the piezometer which is set on the venturi, the water level or hydraulic gradient will drop to 3, then rise again, at the end of the meter, up to within a small distance below its former position (this distance representing the "loss of head" due the whole apparatus), then will run parallel to its former position, as shown at 3 on the down-stream piezometers. The experiments have shown that the velocity of the water, or the discharge through the narrowest section of the meter through the "venturi," is that due the head on the venturi (as represented by the difference in level of two points in the "hydraulic gradient," one taken just above the meter and the other at the venturi), with a co-efficient, which is remarkably constant, whether applied to a rough meter and for a 9-foot pipe, or to a smooth one for a 1-foot pipe, and for all velocities through the pipe ranging from ½ to six feet per second. If this co-efficient is taken without further measurement at 98 per cent., we may be certain from experiments so far made that we shall rarely be over 2 per cent. out of the way. Going back a little, let us take now the other case of submergence, originally represented by the hydraulic gradients aa and bb. It is plain that the water level in the piezometer which is set on the

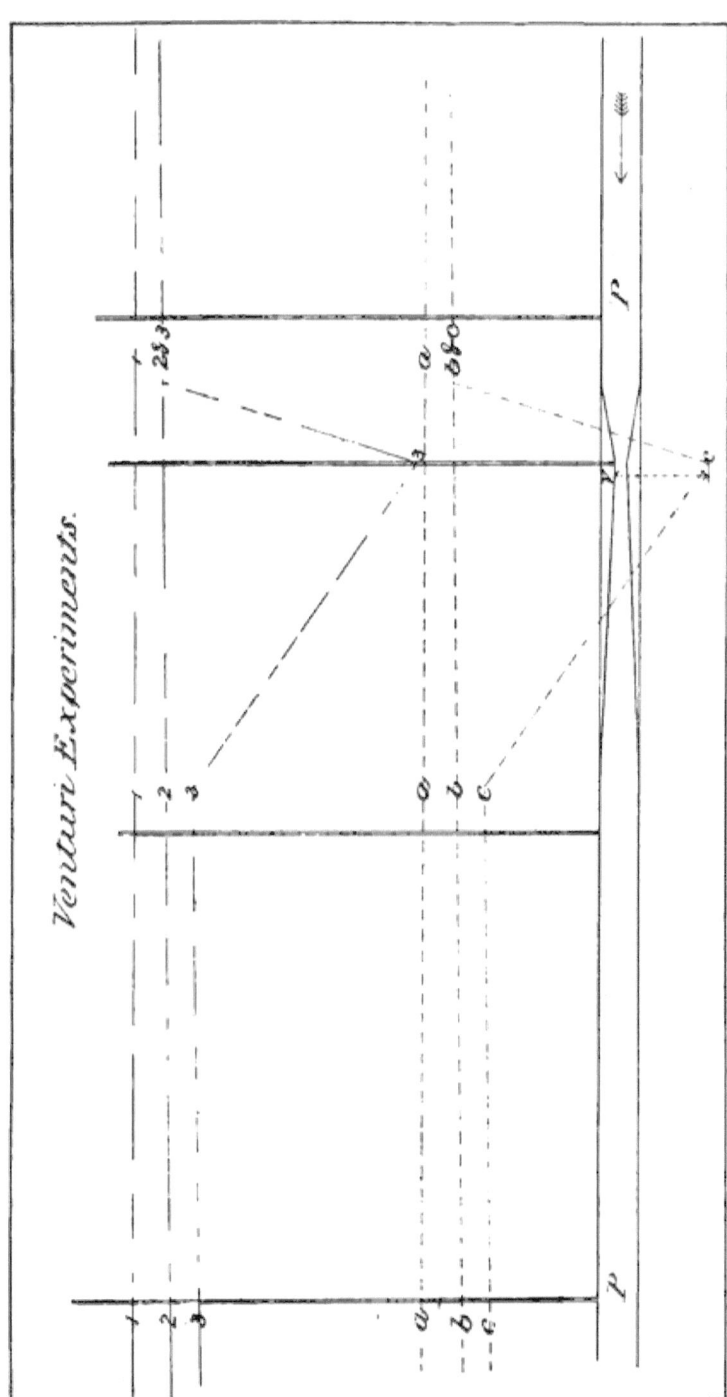

Venturi Experiments.

PLATE XL.

venturi cannot fall below the surface of the stream spouting through the venturi.

But so much of the "depression" in the hydraulic gradient at this point due the velocity of the water, which lies below the surface of the stream just named, will be indicated or exhibited by the sucking action or aspiration or "vacuum" spoken of in the paper; and in measures of a column of water lifted, will exactly equal that portion of the "depression," as shown in Plate XL (or as it may be computed), which lies below the surface of the stream spouting through the venturi. In Plate XL it is equal to the distance v c, indicated by the reference marks.

This is the case in which some form of vacuum-gauge is necessary at the venturi, when separate gauges are used at the up-stream piezometer and at the venturi, as was done during the experiments related in the paper. No such complication is necessary, however, in practice. As the measure sought is the difference of pressure immediately above the meter and at the venturi, a single pressure-gauge suffices. The logical possibilities, depending on the degree of submergence of, and velocity through the pipe, are three, and are exhibited by the table:

	Above the Meter.	At the Venturi.
1	Pressure.	Pressure.
2	Pressure.	Vacuum.
3	Vacuum.	Vacuum.

But in any event there will be a "head on the venturi," or, what is the same thing, Column 1 minus Column 2 of the tabular quantities will always be positive, and will indicate pressure, and may be measured by a single-pressure gauge.

Answering question which relate to the temperature of the water during the experiments recorded in the paper,

I will state that this varied from 67 to 71 degrees Fahr. during the June experiments, with the temperature of the air in the wheel-pit, and of the water in the tub of the pressure-gauge, varying from 66 to 71 degrees.

During the October experiments the temperature of the water was 57-57.5 degrees.

It is undoubtedly true that all hydraulic formulas are affected by the temperature of the water, when that temperature passes beyond those ordinarily found in running water; but in ordinary practice, and without reference to water artificially heated, and as it is at times found in steam engineering practice, no account has ever been or need be taken of temperature that the author knows of.

The formula used in the computations, it will be observed, supposes a discharge through an orifice, under pressures crudely represented by the hydraulic gradient.

A more scientific way would have been to make use of the hydraulic principle first enunciated by Dubuat, but disputed by Navier and others, that the pressure against any point in the walls of any vessel or pipe is always equal to that of the contained fluid, supposed to be in a state of rest, less the height due the velocity past that point.

Or, passing to algebraic symbols, if

P be the pressure in terms of the height of a column of water at the point P, Plate XL, and

P_1 " " at the point P' of Plate XL; if

v " velocity" " P, and

v_1 " " " " P'; also

P_s " pressure if the water be supposed to be still; then

$$P = P_s \frac{v^2}{2\,g}$$

$$P_1 = P_s \frac{v_1^2}{2\,g}$$

and $v_1 = 9\,v$, from the construction of the meter.

Subtracting the equations, we have :

$$P - P_1 = \frac{v_1^2}{2g} - \frac{v^2}{2g} = \frac{80}{81}\frac{v_1^2}{2g}$$

But $P - P_1$ is what we have called Hv, the " head on the venturi."   Or $v_1 =$ the velocity through the venturi,

$$= \sqrt{\frac{81}{80}}\sqrt{2\,g\,Hv} = 1.0062\,\sqrt{2\,g\,Hv};$$

where the first supposition has supposed $v_1 = \sqrt{2\,g\,Hv}$, abstracting in both instances from the co-efficients for actual use, which experiment alone can supply.